Science In a Social CONtext

How Can We Be Sure?

JOAN SOLOMON

ASSOCIATION FOR SCIENCE EDUCATION

BASIL BLACKWELL · PUBLISHER

Acknowledgements

Joan Solomon would like to thank many friends and pupils for their invaluable help and comments, and particularly the many teachers who participated in the SISCON-in-Schools project and tried out these materials in their schools and colleges.

The publishers would like to thank the following for permission to reproduce illustrations on pages: 23 David Austin, *The New Scientist*; 16 John Cleare, Mountain Camera; cover *Oxford Mail*; 16 *inset* Ann Ronan Picture Library; 36 Laurie Sparham, International Freelance Library Ltd.

First published 1983
Basil Blackwell Publisher Limited
108 Cowley Road
Oxford
OX4 1JF
and
The Association for Science Education
College Lane
Hatfield
Hertfordshire
AL10 9AA

ISBN 0 631 91970 8

Typesetting by Oxford Verbatim Limited
Printed in Great Britain

Contents

Introduction

This book is about the nature of scientific explanations, and about trying to make up our minds on scientific and technological matters that affect us. First we consider what it is possible to be certain about and the universal statements upon which arguments can be based. Next we look into the history of science to find out how scientific theories are invented; this includes experiments, prediction and imagination, even controversy between rival theories.

In the last part we consider contemporary issues where there are disagreements between scientific experts about the effects of technology and science. So we shall need to know how we can express our views on these important social matters.

1 Why We Need to Know

It is important for us to try to find out just how far scientific knowledge is true and certain. This is not a matter of philosophy which only concerns the scientists themselves. Whenever problems arise because there has been a public outcry about some disaster, or because a new technology is about to affect our way of living, *scientific experts* are called upon to deliver their opinion. We want some reliable advice. Almost all the government departments employ scientific advisers for the same reasons. It would be comforting if such expert opinion was always entirely sure but, alas, it is only too obvious that experts often disagree!

How do such disagreements come about?

> Are scientific theories themselves reliable and quite certain?
>
> Can experiments decide clearly, once and for all, what is correct in science and what is not?
>
> Could the scientific opinions of experts reflect some other opinions that they hold?
>
> Is it reasonable to argue with the experts if we think that some new advance in technology may affect us? Who knows best?
>
> What can or should we do when science and technology are applied to our society? How can citizens present their opinions and participate in decision-making?

Advertisements often proclaim that 'science has proved . . .', or that some fact is 'scientifically certain'. We may feel a little doubtful about this claim if it is just being used to encourage us to buy a certain brand of toothpaste. Legislation now exists to prevent advertisers telling downright lies in their advertisements. As consumers we need to be able to rely upon the information given about goods for sale.

Government policy about energy, the health service, the environment, defence and scores of situations which will affect us all, has to be made ahead of time. The government consults its specially appointed scientific advisers and often asks them to predict what may happen in the future. Will some type of pollution build up to dangerous proportions? How much power will the nation need and what kind of power stations will prove least dangerous? They will want to know if a new invention should be backed up with a lot of public money from the taxes we pay, because it is likely to prove a great commercial success. Prediction is difficult but the experts are *not* supposed to be guessing. We expect them to use 'proved and certain' scientific theories when they offer their advice.

Our first step, then, will be to find out the answers to some questions about these scientific theories.

> Are scientific statements absolutely true?
>
> Is any kind of knowledge absolutely true?
>
> How do scientists arrive at their theories?

HOW LOGIC BEGAN

The half-legendary Pythagoras is said to have founded a religious mathematical community, including both men and women, in about 580 BC. He left no written works and it is impossible to tell whether even the famous theorem which we all learn in school was really his discovery. There is, however, a saying which is attributed to him – 'a theorem is a platform from which to go higher'.

This is interesting. It means that once you have established one point (for example, that the base angles of an isosceles triangle are equal), you can use this to prove another point (that the bisector of the top angle is at right angles to the base) and so on. The argument leads from one level to another, building on what has gone before. As with the construction of a pile of bricks it all depends on a secure *basis* and a sound *method* of building.

It was not just in mathematics that the Greeks used argument to make points. They were skilled at discussion and debate. The name of Socrates is for ever associated with the kind of verbal enquiry which begins from apparently simple statements and goes on to show what must follow from them. This method of reasoning, from one point to another is called *logic*; it has its own rules, rather like mathematics, and when it is well done it can be totally convincing and certain.

THE RULES OF DEDUCTION

The philosopher who first examined and wrote down the rules of logic was the great Aristotle, who lived in Athens during and after the time of Socrates and probably knew him well. Like all the educated Greeks of his time, Aristotle was well versed in mathematics; the way he analysed logical argument would now be regarded as a part of simple set theory and illustrated by Venn diagrams. He took the simplest form of argument, consisting of three lines only. His most famous example runs:

> All men are mortal
> Socrates is a man
> So Socrates is mortal

Notice that each line has a distinct and separate purpose.

Line 1   This is a universal statement about all the members of a set. It is called a *premiss*. Everyone has to agree to this before the argument can proceed.

Line 2   This is a simpler statement about one member of the set mentioned in the premiss. It is a particular case.

Line 3   This is the conclusion. If you agree to the first two propositions you are bound to agree to this, the rules of mathematics or logic force you to.

Venn diagram showing Socrates' logical method.

Such an argument is obviously correct or *valid*. Aristotle examined many such forms of argument in great detail and it has been said that his work was so complete that no advance was made upon it for more than two thousand years – until the twentieth century.

Examine the following premisses and see if you can construct valid arguments with them. You are left to add two statements of your own choice to each premiss, which could logically follow according to Socrates' three line argument

1 All teachers make mistakes. I am a teacher. (Therefore . . .)
2 Some pupils are stupid.
3 No examinations are worth taking.

Whether or not you agree with these statements you will be able to deduce conclusions from them in the *first and last cases only*. The second premiss may well be true but it is not general enough to form the basis of useful deductions. No doubt you will have little difficulty in thinking of a stupid pupil but you could never logically deduce that anyone was from such a weak premiss!

WHEN WE DO AGREE

The difficulty with obtaining knowledge by means of deduction is not the making of a valid, correct argument – that is easy enough to check with a Venn diagram – it is deciding on a premiss that is the problem. A statement is required that is both *universal* and *acceptable* to everyone, and that is a combination which is hard to find in daily life except in special circumstances.

Mathematics

Every equation is a kind of premiss.

$y = 2x$ (or every y is twice as much as the corresponding value of x).
$y = 4$
$\therefore x = 2$

If you are set such an equation, do you argue about its truth? Of course not; why bother? It is an agreed starting point for the problem.

Games

　　'Anyone who steps on the lines is out.'

We all agree to the rules when we set out to play, and the game follows from them.

Legal systems and constitutions

Here is a situation which is altogether more serious and complex. The laws which are enacted by an elected majority in parliament are put into action in the courts of law for all the population. They may serve as a kind of premiss; but there is a range of penalties for breaking them which shows that other factors, special pleas of misunderstanding and personal circumstances, are also taken into account. There may also be those among the population who do not agree with the fundamental premisses underlying such laws. Out of such disagreements grow movements to change the law.

Even such statements as: 'Every man is innocent until proved guilty' or 'All pedestrians have right of way' can be difficult to apply in some situations.

The American Declaration of Independence set out a series of premisses such as:

> 'We hold these truths to be self-evident that all men are created equal, that they are endowed by their Creator with certain inalienable rights, that among these are life, liberty and the pursuit of happiness.' (1776)

After the Revolution these ideas were put into legal form in the American Constitution. Now any citizen, who feels that his 'inalienable rights' are being threatened in some way, can appeal in a court of law.

Religious systems

To a fundamentalist, belief in such a system may involve total acceptance of certain premisses and all that follows logically from them.

　　Christ died to save sinners

　　All men are sinners

For others, religious belief is a far looser system involving more of an attitude towards life than a rigid code of dogma. Religious discussion can therefore be very different from a logical argument.

Rules for classification

Biological classification started with the rules and definitions laid down by the great eighteenth-century biologist, Linnaeus. He, for example, defined a mammal as a warm-blooded animal which always brought forth its young alive. The rules have been changed a little since that time, by agreement between the biologists, and most of us are now taught them at school.

'All warm-blooded animals which suckle their young are mammals.'

'All animals with backbones are vertebrates.'

These statements are agreed and useful for classifying, but they are hardly more than working definitions of names. However we may be sure that biologists would not have bothered to make up such definitions unless they knew, by observation, that such animals did commonly exist.

2 Not So Sure

Here we are on more difficult ground. We have our own experiences, we make observations, and then we like to form opinions and generalisations to apply in every case.

> All red-haired girls have hot tempers.
>
> No mammals lay eggs.
>
> Whenever there is a red sky at night there will be good weather the next day.

Could there be exceptions to these rules? Perhaps you may know of one in the first case, and that will destroy your confidence in this rule. Linnaeus used the second rule as part of his classification system, but at that time the duck-billed platypus had not been discovered – a warm-blooded Australian animal which lays eggs, hatches them, and then suckles its young. For years naturalists believed that it could be no more than a legend until one was found in the act of laying. Now we have to classify the platypus as a mammal, so the rule that no mammals lay eggs becomes no more than a generalisation with exceptions, and not very useful. Folklore about the weather is much the same. It can be useful at times but often lets us down.

Making up rough and ready generalisations does *not* give us agreed premisses and it is not *science* either.

INDUCTION – DOES IT WORK?

In the early days of science it was thought that making up scientific theories was rather like making generalisations from what you observed, and then adding some reason for it. Scientific theories often sound like universal premisses.

> 'All objects in the universe attract each other with the force of gravity.'

Is it just a generalisation? Not every object can possibly have been examined so scientists cannot be certain. But why do they want to use this idea of gravity? You will probably guess, quite rightly, that it is to *explain* why things fall downwards, why the moon goes round the earth, why the planets go round the sun, and why the stars rotate in their galaxies.

Guessing the *cause* which makes things happen is the business of science; it is challenging and difficult. Unlike deduction it does *not* start from an agreed universal premiss, but tries to make up a universal theory to explain the various pieces of evidence. This is called *induction*, and it is very far from certain.

Aristotle, who was more of a biologist than a physicist, tended to under-rate the difficulty. For example, in discussing the universal cause of thunder, he said:

'Thunder occurs in clouds (Indeed would we call it "thunder" if it didn't?).'

Then he said he saw, by the action of 'quick wit', the missing link:

'All thunderous noises are made by the quenching of fire. Therefore fire is quenched in a cloud whenever thunder sounds.'
What is wrong with that?

The next philosopher to examine induction, or 'scientific method' as it is sometimes called, lived at the time of the Scientific Renaissance in England. This was Francis Bacon (1561–1626). He was a great enthusiast for science and was largely responsible for the founding of the Royal Society. Bacon thought he could see where Aristotle had gone wrong and decided that far more evidence (observation and experiment) was required before a 'cause' could be established. He set himself the problem 'What is the nature of heat?' and proceeded to write down long lists of things that were hot and things that were cold and even suggested some experiments. Such phenomena as rising convection currents, expansion, and friction stood out clearly from his lists so he wrote the conclusion: 'Heat is expansive upward motion'.

These conclusions are obviously incorrect but it must be admitted that Bacon's method seems more scientific than Aristotle's quick guess; certainly it was more thorough an examination. He could not have known about helium balloons, used commonly nowadays for meteorological purposes, which rise high into the atmosphere, expand to many times their original volume and yet cool to a much lower temperature.

'Finding fossilized bones of arctic animals in the tropics indicates either climatic upheaval, continental drift or that paleolithic man had a zoo there.'

By the eighteenth century philosophers like David Hume had shown quite clearly that induction could never be made to give certain answers in the way that deduction did. He made two important points:

1   Finding out that one event accompanied another does not necessarily prove that the first caused the other (such as rain and thunder).
2   A list of instances can never be complete so a universal conclusion cannot be logically inferred from it (such as Bacon's attempt with heat).

Observations and experiments give clues which may suggest a new theory in science. This is induction about which we cannot be sure. But it is possible to use the rules of logic, not to establish the correctness of the theory, but to *deduce what must follow* from it. A scientific theory is

14

always a grand universal statement, which includes an explanation for why things happen. So it can also make *predictions*. For example:

'All putrefaction is caused by airborne organisms.'

This theory had originated with an Italian biologist in the eighteenth century. He had shown that if you sterilised the contents of a tube by boiling it, and sealed it so that no air could get in, then no microscopic life would be found in its contents. Some biologists, like Louis Pasteur, believed that this showed that these organisms were carried by the air. Other scientists of the time still held that micro-organisms arose spontaneously in the rotten food and that all the air did was to help them to grow. (For centuries people had believed that maggots, the larvae of flies, were actually created in putrifying meat, and even that crocodiles were generated from the mud of the Nile!)

There was one experiment and two possible theories to explain it. Then Pasteur made a new prediction from his theory:

Air from remote mountain regions is unlikely to contain living organisms. *Therefore* putrefaction is not likely to be caused by such air.

To test this prediction Pasteur first prepared 73 specially blown glass vessels containing a clear nutritious liquid which had been sterilised by boiling. Each had a long swan-like neck which was sealed up to prevent air getting in. Then Pasteur and various scientific assistants, a baggage-load of equipment and special guides, set off for the Jura mountains.

When they were well away from human habitation, in the foothills of the mountains, twenty of the special flasks were opened by breaking off the sealed glass at the end of the neck, and then sealed up again with a flame. Within a few days the liquid in eight of the flasks began to ferment and go cloudy.

Then the scientific cavalcade, with their portable laboratory equipment, began climbing up the mountains. At 850 metres they stopped again and exposed twenty more of the flasks to the mountain air. This time only five of the flasks went cloudy. Finally they went up into the high Alps reaching the great glacier, Mer de Glace, on Mont Blanc itself. Here, surrounded by ice and snow, carefully holding the flasks at arm's length away from his breath and clothing, Pasteur let the air into another set of twenty flasks before sealing them up again. In only one of these flasks did a slight cloudiness appear.

The prediction had been fulfilled and, after three years of public argument, the scientists of his day allowed themselves to be convinced of the theory that Pasteur had championed in such a spectacular way.

The Mer de Glace on Mont Blanc. Pasteur believed that air from remote mountains was unlikely to contain living organisms and therefore putrefaction was not likely to occur. He tested his theory, successfully, on this glacier. The inset shows one of the flasks which he used in this experiment.

This new scientific understanding of the existence of airborne organisms was to prove of enormous practical benefit, both in medicine for the production of sterile conditions, and in the French wine-making industry.

So we see that logic can be used to make predictions. To find out more about the making of the theory itself we cannot do better than to examine the creation of some actual scientific explanations.

3 The Birth of Scientific Theories

The rainbow is an inspiring source of legend, a beautiful enigma and also a very early object of scientific study. By the thirteenth century it had already become obvious that the clue to its mystery should be sought in experiments which traced the paths of light rays through water.

Most people know that there are two bows which can only be seen with the back to the sun. Fewer know that the angular sizes of these bows are always the same, 42° for the stronger primary bow and 51° for the fainter reversed secondary bow outside it.

From a mountain in the evening or early morning much more of the circle can be seen. From an aeroplane it can occasionally appear complete. The drops of water are essential. So apparatus of various shapes was designed to trace light rays and measure the angles at which they were bent as they passed into and out of water. Gradually the data became more accurate and yet no reliable theory emerged as the

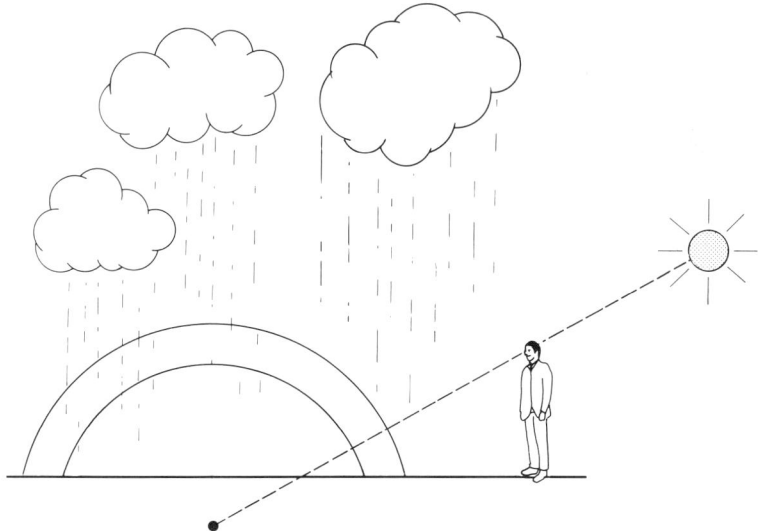

The rainbow: an observer sees the bow when his back is to the sun and the centre of the bow is below the horizon.

17

centuries passed. Mere juggling with the figures could not produce a generalisation, let alone an explanation of the angular sizes of the bows.

The breakthrough occurred in the 1620s. The French philosopher René Descartes was watching cannon balls being fired across a river and noticed the change of direction when they hit the water. The reason seemed obvious, they travelled slower in water than in air.

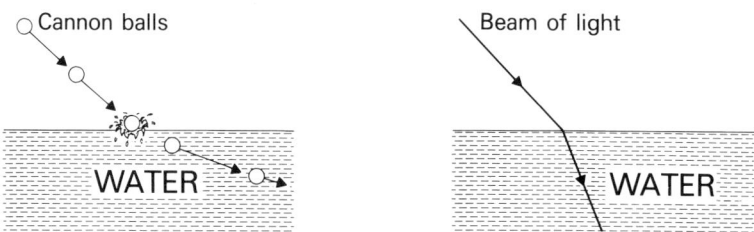

Why do light rays bend?

Descartes made the analogy: light rays were like a stream of tiny particles (mini-cannon balls) which speeded up instead of slowing down when they entered water. For the first time someone had made a guess about *how* light moved and *why* it changed direction. (Ignore the fact that this is not now the accepted theory of light).

Unlike Bacon's sterile lists, Descartes' theory proved fertile. From it he could deduce a numerical law of refraction and compare it with the experimental results. They fitted almost perfectly. Then he could plot the paths of rays of light through a drop of water and calculate the angular size of the two rainbows from his theory. This gave exactly 42° and 51°. The theory was also taken up by Isaac Newton who used it to explain other optical effects. For nearly two hundred years it was a successful scientific theory.

Now ask yourselves these questions:

1   Was there any direct evidence that light was a stream of tiny particles?
2   Why could this theory predict so many other results while most generalisations do not?
3   Did it arise directly from experiment and, if not, what was its relation to experiment?

The birth of a new scientific theory need not be quite so sudden and complete. The theory of Continental Drift, for example, has been around for a long time – first suggested, in all probability, by the curious fit between the maps of Africa and South America. In the 1920s it enjoyed increased attention due to the enthusiasm of Alfred Wegener, a German meteorologist. He collected a great deal of new evidence ranging from the age and type of rocks on the two continents to the distribution of earthworms.

The same genera of worms are still to be found in Australia as in India, in South Africa as in South America, in Central Africa as in Central America. As worms could not have swum across the oceans, Wegener thought that this was convincing proof that the continents had once been in contact and had subsequently moved apart. Other scientists, at this time, suggested that the distribution of worms could be explained by imagining that thin 'land-bridges' had once joined the immovable continents together.

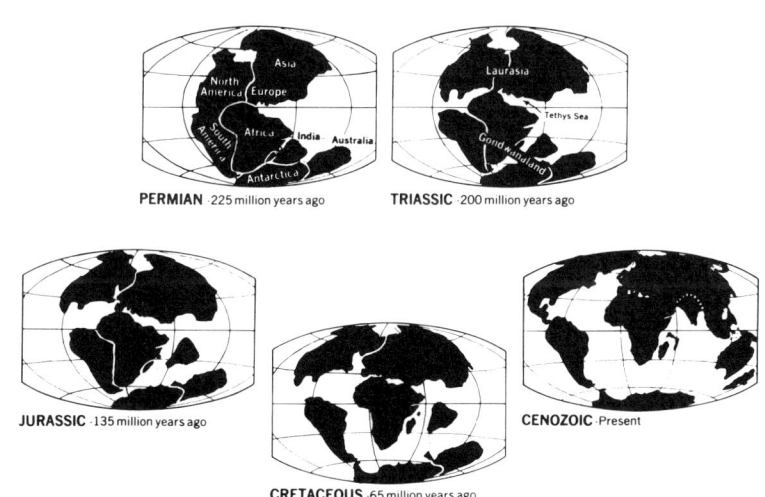

How the continents could have moved apart in 225 million years.

Wegener's theory failed to win over the scientists because there seemed to be no evidence of a force strong enough to move such immense land masses over such vast distances. Geological theory of the time was concerned with mountain building produced by the shrinking and folding of the solid crust of the earth.

19

Wegener died in 1930 and the scientific world remained unconvinced until 1960, when the new science of Plate Tectonics burst upon the scene. It was suggested by the American Harry Hess that a long crack still exists on the floor of the Atlantic Ocean where the two plates are slowly being drawn apart. Molten rock (basalt) rises from below to fill the crack continuously and so creates new crust beneath the ocean.

The complementary action of Plate Tectonics must be the slow consumption of the other edge of the moving plate as it is pushed against a thicker continental crust. Often new mountain ranges (such as the Andes) are formed and the thin oceanic plate may be pushed underneath the boundary of the other plate, to be slowly dissolved in the hot mantle below.

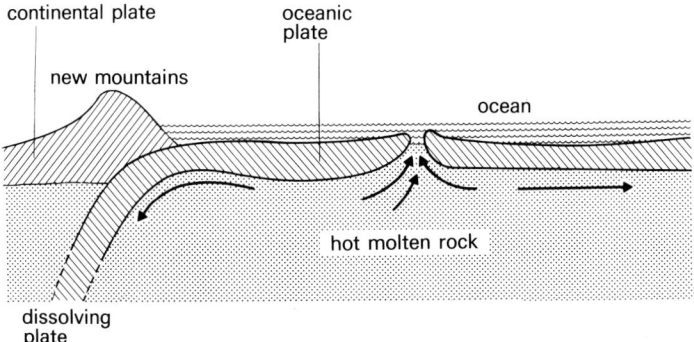

Cross-section of two plates moving apart under the sea.

The effect of this theory on geology has been dramatic. There have been surveys of magnetic field reversals in the new basaltic rock beneath the Atlantic Ocean confirming the symmetrical out-pouring of this volcanic material to the east and the west of this crack. Careful measurement of surface gravity across the interface between oceanic and continental plates on the Western sea-bed of America have shown the angle of descent of the leading egde of the lower slab. Seismic disturbances have traced it to the astonishing depth of 600 km and more below the 70 km crust. A whole new area of investigation and experiment has emerged. As you can see from the diagram there are now new explanations for the rise of mountains in place of the old ones about the shrinking crust.

One of the new investigations, stimulated by this theory, has been an examination of rock magnetism.

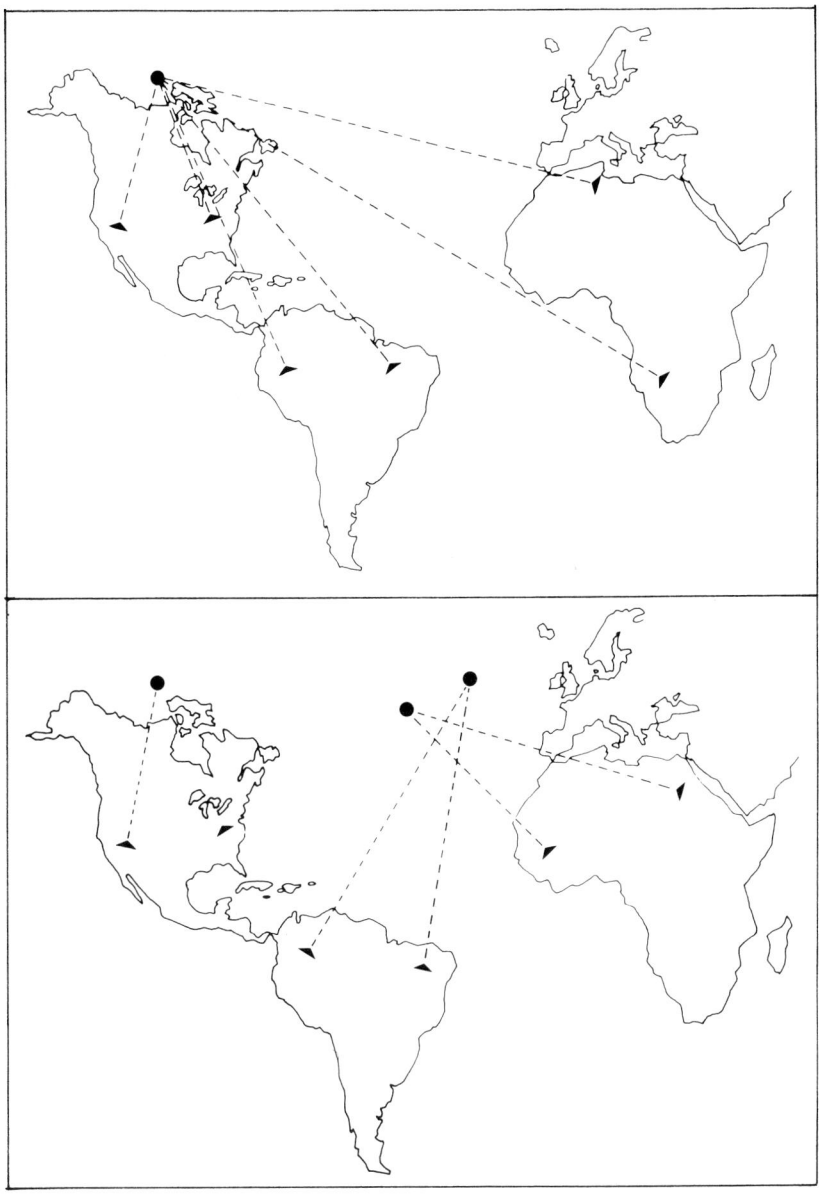

The Americas and Africa: (top) present-day compasses point towards the magnetic North Pole. (above) The evidence of rocks formed 200 million years ago does not correspond to the location of the North Pole today.

Could the continents have moved? Try swivelling them until the locations coincide.

21

For the story of another brilliant scientific discovery you could read *The Double Helix* by the Nobel Prize winner Jim Watson, which tells of the events leading up to the discovery of the structure of DNA from the point of view of a scientific participant.

WHAT SCIENTIFIC THEORIES ARE LIKE

There are at least *four* similarities between these stories from scientific history.

1 Scientific theories go far *beyond the observable effects* and speak about particles too small to see, or huge movements too slow to feel.
2 They suggest a *mechanism* for the effect, i.e., how it happens.
3 They stimulate *new experiments* by making *predictions* about them.
4 They can *change*.

During 1981 a famous trial took place about the teaching of creation in American schools, and it became essential to establish what was a scientific theory and what was not. By the terms of the American constitution religious beliefs may not be taught in school. (This is not the case in Britain, but when you consider how many of the early settlers fled to America there to escape from persecution and to gain the freedom to raise children in their own religious beliefs, it is not at all surprising.) Nevertheless many religious people wanted the biblical story of creation, as it is told in Genesis, taught to their children either instead, or as well as, the theory of evolution. A law had been passed in the previous year which laid down that 'creation-science' should be taught *during science lessons* and given equal time, and the same balanced treatment as was given to Darwin's theory of Evolution.

Then the Evolutionists fought back. As you can see from the cartoon, which is grossly unfair to the Creationists, feelings ran high. The scientists prepared their case carefully and several famous professors were called to describe how rocks were dated and the results of experiments on generations of fruit flies. The very first witness that the Evolutionists brought was not a scientist, but a philosopher.

The Creationists had based their claim for equal and balanced teaching time on their assertion that Creationism was a science like Evolution. The task of the philosopher was to show, from the writings of the

22

IF YOU-ALL SAY WE STARTED OFF
SWINGING IN THE TREES YOU COULD
END UP SWINGING FROM ONE.

Creationists themselves, that it was not. He picked out these sentences from their most popular book to argue his case.

> 'We do not know how God created, what processes He used, for God used processes which are not now operating anywhere in the natural universe. This is why we refer to divine creation as special creation. We cannot discover by scientific investigation anything about creative processes. . .'

At the end of the trial, in January 1982, the judge ruled that the 'equal time' law was indeed a breach of the American constitution and, for the State of Arkansas at least, it was repealed.

Take the four points given on page 22 and try to show how three of them do not fit the Creationist's position. They do not prove that anyone is right or wrong, only that creationism is not science.

4 Learning Theories and Explaining Experiments

The more known effects a theory can explain and the greater its capacity to predict new ones, the more convincing it is. Really successful theories tend to 'snowball' because they are taught at school and university so thoroughly that the scientists begin to use them like a pair of mental spectacles whenever they design or perform experiments. This is inevitable.

For example, we are all taught that gases consist of millions of fast-moving molecules. We get trained to look at inflated balloons and imagine a chaotic host of molecules within which are bombarding the side of the rubber so hard that they distend it. In reality we never see these molecules. The nearest we ever come to it is watching 'Brownian Motion' in a capsule of smoke under the microscope. The tiny smoke particles seem to be jostled about and we call this evidence for the existence of bombarding molecules. Now Robert Brown after whom the phenomenon is named, lived before the kinetic theory of gases had become established. He watched the movement of pollen grains in water, a new effect that no scientist had reported seeing before. The grains appeared to wander about, growing and shrinking as they went. (He was not too surprised about this: pollen is the agent of reproduction and hence the source of life itself.) When he discovered that even powdered minerals showed the same motion, Brown still attributed it to the live force of 'organic' molecules within them. *The observation can depend upon what theory the scientist has in mind.*

An experiment on you

The *objects* you will be observing in this experiment are pollen grains which are quite as small as bacteria or smoke particles, and come from living plants. However the real *subject* of the experiment is not pollen, but your combined powers of observation, imagination and explanation.

Pollen can be obtained fairly easily in the late spring or summer months. For this experiment you need to use the very small pollen grains found in wind pollinated plants, like grasses and trees. (The pollen from insect-pollinated plants is too large.) Take a clean, dry microscope slide and carefully place one drop of water on it. Then, with a paintbrush, gently dust a little pollen off the flowering part of your specimen, e.g., a catkin, so that a film of pollen may now cover the surface of the drop of water.

Place the slide, without any coverslip, under the microscope. You will need to use the high-power lens, but make sure that it does not touch the water surface. Now shine a narrow beam of light onto the drop of water from the side. Focus the microscope carefully until you can just see tiny specks of light against a dark background – these are the pollen grains. Watch them for two or three minutes.

1 Describe their movement.
2 Are they trying to cluster together or move apart? Are they swimming around or moving in a random jerky way? Are they keeping together in a stream? Brownian motion or live movement?
3 Does your explanation influence how you see the movements of the pollen?
4 Compare notes with the others. Do all the observations agree? Does each observation agree with its explanation?

(As it might influence your observation this book will not tell you whether the accepted view of this movement is that pollen grains use some unseen flagellae for swimming, or that they are being subjected to Brownian motion from the bombarding water molecules.)

Scientific theories often speak about unobservable things – high speed molecules, electrons inside cables, viruses causing disease and an expanding universe. We are first taught the theory and if we absorb it well into our imagination it will explain a whole host of experiments and add valuable detail to the theoretical picture. Without it the experiments themselves might prove almost nothing!

Are we being brain-washed?

Can such a theory ever be proved wrong?

HOW WE LEARN SCIENCE AT SCHOOL

If you were lucky you learnt science in a well-equipped laboratory where you were able to do experiments. Young children enjoy messing about with apparatus, looking through microscopes, burning things with a Bunsen, and lighting up lamps with electricity. To their science teacher, however, the real object of the lesson is to teach scientific theories. They want to teach children to 'see' the cells in plants and animals, to understand that oxygen from the air is being used up during burning, and that an invisible current of electricity flows through the wires and round the circuit whenever the lamps light up. Usually the teacher explains the theory behind each observation either before the experiment is done, or immediately afterwards. Then the observations that the pupils make become evidence for the theory.

Do you think that the pupils could 'discover' these theories for themselves by just looking?

Did you find school experiments surprising, convincing, or boring?

How would you try to convince a child of 8 or 9 years old that air was not just emptiness and that electricity 'flows'?

5   Science Changes its Mind

Theories do fall and when they do it is a dramatic event in science, like a revolution, since it runs counter to everyone's trained expectations. There are advocates and adversaries, evidence and prejudice; but above all there are new experiments and new predictions. We will examine how the theory of light underwent such a revolution.

By the beginning of the eighteenth century there were two rival theories of light. One of them, based on the work of Descartes and Newton, held that rays of light were streams of tiny particles. The other, due to Christian Huygens, who worked in France, held that light was a compression wave – rather like sound – which travelled through an invisible weightless substance called 'aether' which permeated most substances and filled space itself.

Both theories could successfully explain the laws of reflection and refraction, even though they were so different.

Reflection

Light particles bounce off the mirror

Light waves are reversed at the surface of the mirror

Mirror

Mirror

Particle theory

Wave theory

27

Light is *refracted* into glass because the particles are attracted into it (this makes them move *faster*.)

Particle theory

Light is *refracted* into glass because this *slows* down the wave and twists round its direction of movement.

Wave theory

Notice that the two theories made opposite predictions about the speed of light in glass. Unfortunately a direct measurement of this speed could not be carried out with the apparatus available at the time, so there was no easy way of judging between them.

Both theories had difficulty in explaining some other effects of light but, on the whole, the honours seemed fairly even and either might have been accepted.

In practice it was the particle theory that became established by common scientific consent. In England it had always been the favourite partly because of a suspicion of this 'intangible aether' and partly, perhaps, through a rather disreputable patriotism. Isaac Newton had been an illustrious Englishman and even a century after his death when his theory was being challenged, a letter was written to *The Times* suggesting that his scientific adversary be 'black-balled' from all the best clubs in London!

In the early years of the nineteenth century experiments were carried out by Thomas Young in England and, slightly later, by Augustine Fresnel in France, to show that two rays of light could 'interfere' to produce patches of darkness.

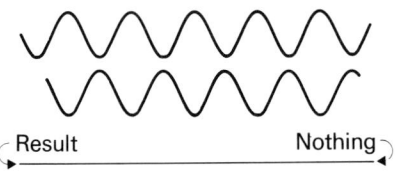

How waves interfere: the two waves are 'out of step'. The crests of one coincide with the troughs of the other cancelling each other out and leaving no wave at all and so no light.

28

This is completely understandable on any *wave theory*. If two waves are 'out of step', like the two shown, and they overlap the crests will superimpose on the top of the troughs. These opposite effects will neutralise each other and the wave will be 'neutralised'. The effect can be easily seen with water waves but the experiment with light waves was quite awkward to perform.

The *particle theory* was helpless to explain the effect. Two light particles should always give twice the light – they could not be 'out of step'. It seemed as if the wave theory had already won the day!

Nevertheless there was a time lag. The scientists had all been brought up to 'see' light rays as streams of tiny bullets and old habits die hard. They did not want to change. In 1817 the French Academy launched an open competition to find an explanation of these interference effects on the traditional particle theory – but it could not be done. Fresnel's paper which used the new wave theory of light was reluctantly awarded the prize.

From then – until the next revolution in optics! – the wave theory became the accepted view of science.

WHAT IS BURNING?

When something burns – wood, paper, even magnesium – smoke and flame usually come off it leaving an ash which is very light. For hundreds of years scientists believed that substances contain a part, they called it 'phlogiston', which escaped from it when it burnt. This seemed a sensible enough theory, even though it was known that a few metals such as lead and mercury actually became heavier when burnt. For a while during the eighteenth century two rival theories of burning existed.

Phlogiston theory

Charcoal burns away to nothing, so it must be pure phlogiston. Lead and mercury only burn with difficulty so they contain very little phlogiston. If you heat the 'ash' of lead with charcoal it puts back the phlogiston to make lead again.

It was easy to explain why wood got lighter when burnt, but hard to explain why some metals got heavier . . . unless you could believe that phlogiston actually had a 'negative' weight.

Air (or oxygen) theory

Everything takes in something from the air when it burns to form ash and gas. If you heat up the metal ash with charcoal, the air part (oxygen) will be removed so making the metal again.

It was easy to explain why some metals got heavier when they burnt, but harder to explain why wood got lighter . . . unless you could somehow catch the smoke and weigh it.

CHARCOAL + METAL ASH   METAL

Making metals with charcoal. Both the oxygen and the phlogiston theory could explain the result, but in different ways.

In 1772 the French chemist Antoine Lavoisier performed an extremely careful and accurate experiment. He heated mercury in an enclosed space with a measured quantity of air for several days. He then measured the amount of air used up. Carefully scraping the bright orange 'ash' off the mercury, he weighed it and then converted it back into mercury again. In this way he found the *gain in weight of the mercury* when it had first burnt. It was exactly the same as the *weight of air which had been used up*.

The followers of the phlogiston theory could not explain away these careful results. More successful experiments followed. Lavoisier managed to collect pure oxygen which he identified as the part of the air used up during burning. He also showed that hydrogen combined with oxygen to form water when it burnt. The sheer weight of experimental evidence became too much for the phlogiston theory, and the oxygen theory finally won the day.

NOT SURE YET

The process of change and challenge is still going on in science. There are always bound to be some new experimental results which are being explained one way by one group of scientists and another way by others. Whenever a new theory is being proposed we can be sure that it will take time to convince all the scientists of its likelihood. There are arguments going on now about how new species evolve, about particles which may be coming out of the sun, about the rings round the planets and about the causes of cancer.

Sometimes a new and disturbing idea gets into the newspapers and spreads alarm, like the use of aerosols destroying the ozone layer which protects us from radiation, and then scientific opinion changes. It is now thought that aerosols are not doing any harm. As we have seen, it is hard for scientific theories to be sure.

6   When the Scientists Disagree

We began this book with the problem of the disagreeing experts. We can now see reasons why scientific knowledge is not as sure as some people seem to think, but there are two other reasons for the conflict between experts when the subject is possible health hazards.

The first reason is that tests on humans, where danger is involved, cannot even be contemplated. So we are really asking the scientists to guess on the basis of other information. Take the case of atmospheric radiation from the explosion of nuclear weapons. Of course it is deadly in large doses, as when the bomb was dropped on Hiroshima (see *The Atomic Bomb* in this series), however, the two scientists who invented the hydrogen bomb – Edward Teller in America, and Andrei Sakharov in Russia – hold quite different views on the danger of radiation from atmospheric tests.

Sakharov was worried about experiments with flies and mice which showed that there was no lower limit of safety. Every increase in radiation caused some more genetic damage. He felt sure that the same would be true for the human population and he warned the Soviet government that atmospheric testing should stop. They took no notice. Teller based his views on the wide variation in *natural* background radiation to which humans have always been exposed. He felt that too much fuss was being made and that the public's fears were being raised quite needlessly. He even testified in front of the American Senate that 'a little radiation can be good for us'. Natural radiation is around us all the time in small amounts – from the earth and from outer space.

You may suspect that personal factors were influencing Sakharov and Teller. In other cases this can be very much clearer. If you were the proud inventor of a new areoplane engine would you be easy to convince that the extra noise or pollution it produced at take-off was serious enough for the aircraft to be grounded?

For both these reasons, i.e., the impossibility of running dangerous tests on humans and the personal involvement of scientists, it is often quite easy for the 'experts' to disagree. It happens almost daily in the law courts, in the case of damage from buildings, medicines, pollutants etc.

We are not likely to know better than the experts about science but we have every right to express our opinion about the way we want *our society* to use science. First we will need to listen to both sides of the argument from the experts and from others. Then we may feel strongly that one course of action would be better than another. What can we do?

1   We can use our votes in the local and parliamentary elections.

But the topic I am interested in was not mentioned in any of the party manifestos.

One thing you could do is to attend some of the election meetings of the candidates and ask them directly what their view is and what action they would take if elected. Of course the candidate that you vote for may not get in.

2   We can write a letter to our MP. Whatever his party, he represents all of his constituents. We can also get a group together and 'lobby' him at the House of Commons.

What's the use? My MP is bound to vote as his party tells him to.

That is only partially true. There are some social issues, such as the Abortion Bill, where there is a free vote. Your MP will then be quite able to represent the wishes of his constituents, as far as he knows them. It is worth trying.

3   Take a close interest in local politics. If the issue in which you are interested concerns local conditions – a threatened demolition or a possible health hazard – the council will be involved and there may be letters in the local paper. You could write to your councillor or to the paper.

Who's going to listen to me?

That is the point about a democracy. Everyone has a right to be heard but it must be the majority who decide. It may well be that there are a lot of other people who feel like you do. A letter to the local paper or a door-to-door canvassing of opinion finds this out; sometimes a group of local residents gets together to express its views as a street group or a tenants association.

A little time ago a hospital dumping ground, surrounded by a high wire fence, was found to contain some mildly radioactive waste material. The local people formed a group to protest. The hospital produced evidence from an expert that the radiation liberated into the environment was so weak that it could do no harm. The residents were advised by another expert that continual close exposure to such contaminated material might be harmful. They decided that no wire fence or notices could keep out really determined children or rodents; however much it might increase the local rates they paid they wanted the waste removed and buried far from inhabited land. They even got the chance to put their case on television. In the end this 'pressure group' won the day.

Whenever there is a new local project, such as the building of a road, or the construction of a factory, an application for planning permission has to be made. Then the local authority is required to collect views about the project from any local residents or other groups which may be concerned. Sometimes there is so much interest or alarm about the issue that it is decided to hold a public inquiry.

An interesting example of this took place in September 1981 when Derbyshire County Council held an inquiry into the dumping of dioxin wastes in a field near the small village of Stretton. Thirteen years earlier there had been a factory explosion in which this deadly substance had been released, and all the contaminated material was buried – without notifying the council of its exact location. Then, in 1978, a mining company applied for permission to begin operating in this area, very near the place that the villagers believed the dioxin had been dumped.

At the inquiry evidence was given by local residents and by the Water Authority. (The owners of the factory, and the firm which transported and tipped the dangerous wastes, refused to attend. The inquiry had no power to force them to do so.) Expert scientific witnesses, called on behalf of the Water Board, and by the residents' pressure group, gave evidence that dioxin loses half its toxicity in nine months . . . or in twenty years!

Why do you think these experts disagreed?
Do you think this was a moral or legal issue?

The biggest and most important inquiry yet held on a scientific or technological issue was the Windscale Inquiry. In 1977 British Nuclear Fuels Limited applied for permission to build a large new factory at Windscale in Cumbria, where they already operated several other nuclear plants. This factory was to 'reprocess' the highly radioactive waste materials from nuclear power stations, both in Britain and abroad, so as to extract plutonium from them. This plutonium could then be used either to power a new sort of nuclear power station (the Fast Breeder Reactor) or to make nuclear weapons. A public inquiry was set up, opened three months later, and lasted for a hundred days.

As with all such local inquiries, local effects on the residents, workers, environment and employment were considered in detail. There was a lot of anxiety about hazards to health from radioactivity, many expert witnesses were called on behalf of the applicants and the national environmental group Friends of the Earth. Once again they gave conflicting estimates of the dangers involved.

Some of the problems considered were even more far-reaching, having both national and international consequences:

> Was there a need for nuclear power? Should Britain build Fast Breeder reactors, fuelled by plutonium, to meet future energy needs? Such questions are usually matters of government policy, but here they were argued out – shortage of fossil fuels, controlling consumption, risks of nuclear power, alternative energy from sun, wind or waves – in open court.

Windscale demonstration and rally against nuclear power in London, 1978.

Would it be ethical to supply plutonium in this way to countries which might use it to make weapons? In 1970 Britain had signed the Treaty on the Non-Proliferation of Nuclear Weapons, in which we had agreed neither to sell weapons to non-nuclear countries, nor fissionable material, like plutonium – except under special safeguards (which were not thought to be effective).

Would the production of so much plutonium encourage international terrorists to converge on Britain in order to steal it? It was argued that this could require such strict security that it could effect our civil liberties and policing.

In the end the judge and his assessors recommended the giving of planning permission, subject to certain safety precautions. But the questions raised by the inquiry itself have not yet been answered.

1   It is expensive to present a legal case at such an inquiry. Should *anyone* be allowed legal aid in order to give all points of view?
2   One of the BNFL managers was heard to comment 'They weren't just objecting to our plans . . . they want to change society.' Do you think an inquiry is the right or wrong place for such opinions? Could they ever become a new part of our democratic government?

Suggested Reading

Straight and Crooked Thinking   Robert Thouless   (Cambridge University Press)

For those who want to read more about the uses and abuses of logic this book is about the simplest that can be found. Contains some interesting and thought-provoking material.

The Double Helix   James D. Watson   (Weidenfeld & Nicolson)

Lucy. The Beginnings of Humankind   D. C. Johanson & M. Edey. (Granada)

Microbes and Men   Robert Reid   (BBC Publications)

All three of these books are about particular scientific discoveries in a human setting. The first two are written by the discoverers themselves and are all the more interesting for it. Neither are specially easy to read but contain anecdotal accounts of the excitements of scientific research which bring the accounts to life. The third is easier material, well illustrated and with emphasis on the social implications of these discoveries – vaccination, asepsis, fermentation – all written with racy human detail enough to fascinate any 16, 17 or 18 year old with an interest in medicine.

Talking about Government   Eileen Bostock   (Wayland)

This is in the series 'Talking Points'. It is easy to read and well illustrated with photographs and quotations from the politicians and their critics. In particular it addresses the question: How can you influence the government? (15 years and older)